LAIYANAI FOREST
Munge River
NGORONGORO
Lake Makat
Oljoronyuki River
Seneto Stream
CAMPSITE
UNDER FIG TREE
LERAI FOREST
CRATER
MONTAINE HIGH FOREST
0 1 2 3 4 5
SCALE OF MILES

SAFARI TO NGORONGORO

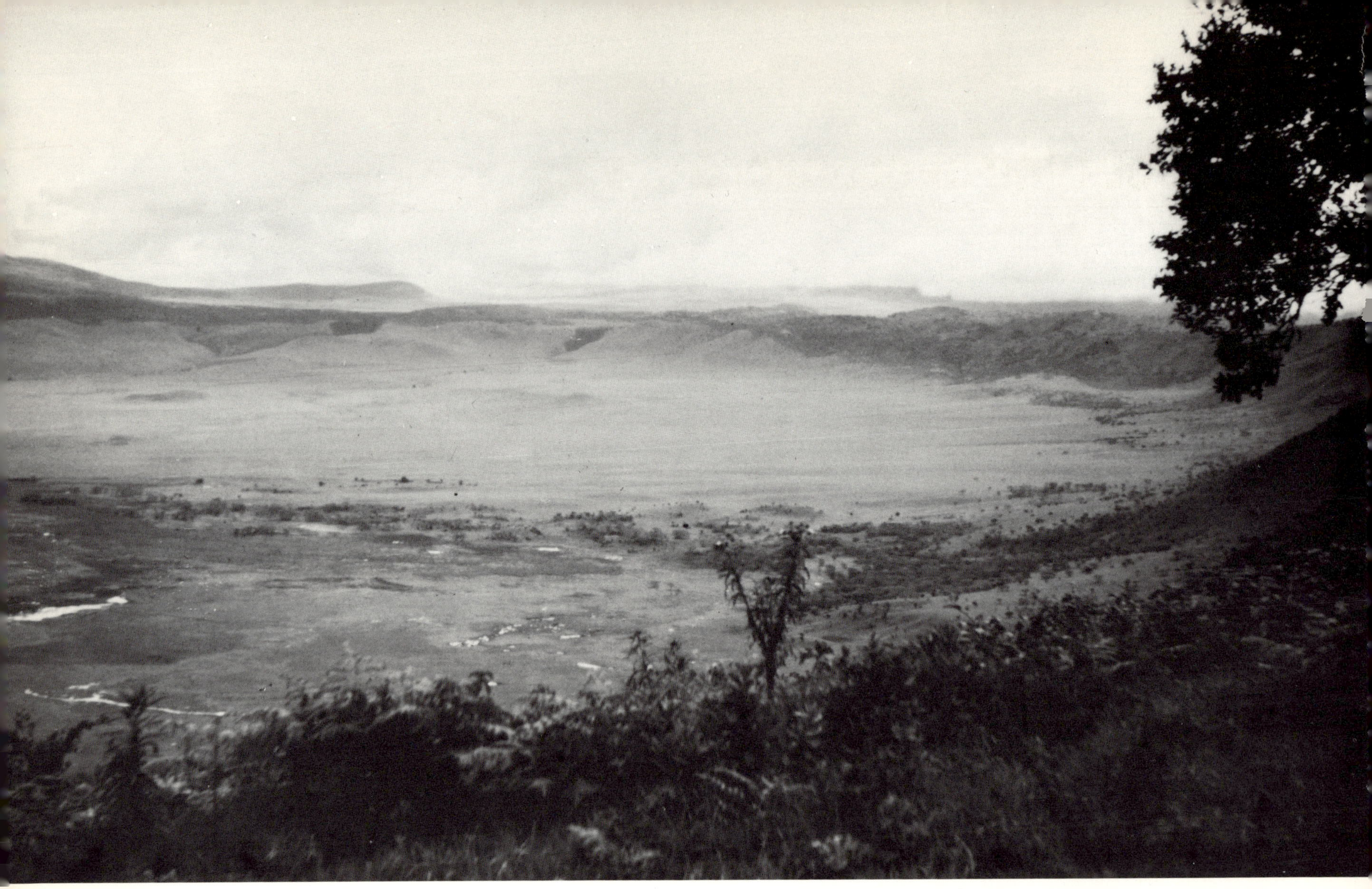

about the ngorongoro crater

Thousands of years ago the Ngorongoro* Crater was an active volcano surrounded by six similar volcanoes. They all erupted. Only one collapsed in the center: Ngorongoro. The others remained tall, conelike mountains.

Over the centuries Ngorongoro became a huge open bowl with forested slopes forming walls 2,000 feet high. The floor of the crater is an open plain—126 flat square miles relieved here and there by conical hills. Gurgling streams flow down the crater walls, irrigating swamps and pools, and a soda lake near the center, called Lake Makat. There are two patches of forest on the floor, Laiyanai and the larger Lerai.

Approximately 25,000 mammals and anywhere from 10,000 to half a million birds live in the crater. Birth and death, courtship and struggle all take place in this wild kingdom. Hunting is outlawed here and the laws are strictly enforced.

Naturalists call Ngorongoro the Eighth Wonder of the World. Scientists refer to it as a caldera rather than a crater because its outer walls collapsed in to form the present inner walls and the floor. *It is the largest unbroken and unflooded caldera in the world.* Ngorongoro is situated below the equator in Tanzania, East Africa.

*pronounced ung-gō″rông-gō′rō

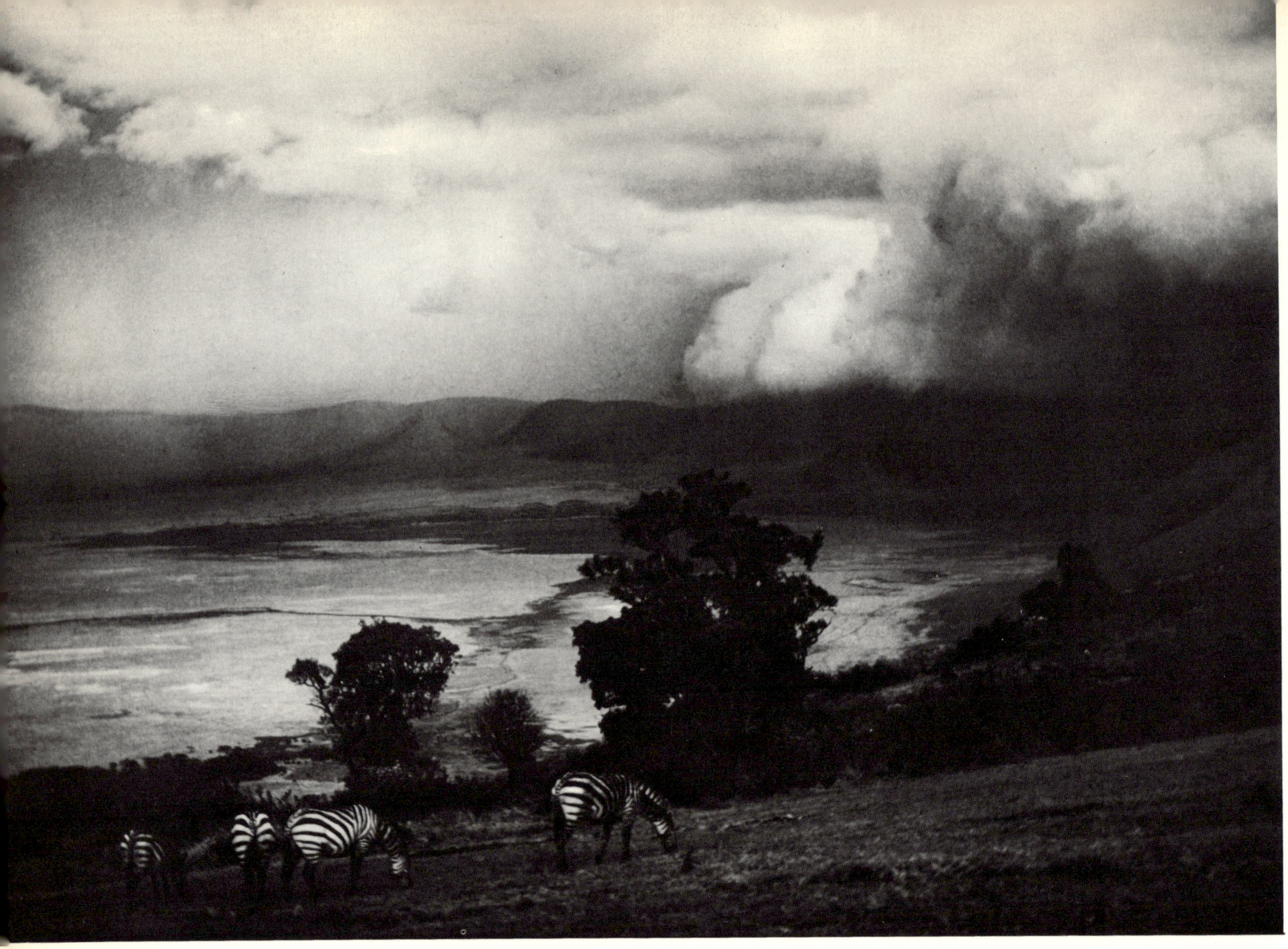

SAFARI TO NGORONGORO

by rachel carr

photos by jan thiede / edward kimball

COWARD, McCANN & GEOGHEGAN / NEW YORK

PHOTO CREDITS

Rachel Carr: page 10 (top)
Edward Kimball: page 10 (bottom), 12 (bottom), 13, 15,
19 (bottom), 20, 22, 24 (top), 25, 27 (top right, bottom),
28, 31 (bottom), 33 (bottom), 34 (bottom), 38, 39 (top)
Dinesh Patel: page 8, 16
Nan Rees: page 2
TWA: page 9, 11, 41
All other photographs were taken by Jan Thiede

Endpapers by Richard Cuffari

Copyright © 1976 by Rachel Carr
All rights reserved. This book, or parts thereof, may not be reproduced
in any form without permission in writing from the publishers.
Published simultaneously in Canada by Longman Canada Limited, Toronto.
SBN: TR-698-20351-8
SBN: GB-698-30603-1
Library of Congress Cataloging in Publication Data
Carr, Rachel E.
Safari to Ngorongoro
Summary: Text and photographs present a survey of the
animals found in Ngorongoro Crater, Tanzania.
1. Zoology—Tanzania—Ngorongoro Crater—Juvenile
literature. (1. Zoology—Tanzania—Ngorongoro Crater)
I. Thiede, Jan. II. Kimball, Edward N. III. Title.
QL337.T3C37 596'.09'67826 75-44152
Printed in the United States of America
Designed by Bobye List

contents

about the ngorongoro crater   3

visiting the maasai   8

our safari continues   12

driving down the crater wall   16

entering the crater floor   18

visiting the lerai forest   25.

birds are everywhere   42

a leopard at sunset   44

camping in the crater   46

wildlife chart   47

index   48

visiting the maasai

The Land-Rover—a sturdy four-wheel drive car—is ready. Our guide takes us through the Montaine High Forest. The land is covered with wild flowers, tall grasses, thick trees and bushes. Excitement builds as we near our first destination: a Maasai village.

"Jambo! Jambo!" a Maasai warrior in his teens calls hello in Kiswahili, a native language of East Africa. He wears a red cotton toga tied in a knot over one shoulder. His body and long braided hair are dyed copper and greased with fat. Dangling earrings hang from his pierced ears. He raises his right hand in a royal salute, striding fearlessly with spear in hand toward his village near the top of the crater. This circular enclosure consists of mud huts made of a framework of poles plastered with cow dung. Livestock are kept safe in the center of this village at night.

A Maasai elder stands proudly near his home. He holds a fly whisk made of silky wildebeest tails. This is a symbol of the elder's authority. Some pretty Maasai girls, adorned with huge beaded collars and colorful bands on their shaven heads, are chanting and dancing. Nearby a young Maasai herder pauses with spear in hand as his family's cattle graze.

Down through the years of wandering from place to place, the Maasai had come to depend on their cattle for their diet. The warriors drink curdled milk with fresh cow's blood as tonic. The blood is skillfully drawn from a vein in the cow's neck. This cut heals quickly. In recent times, the Maasai have added maize porridge, fruit and vegetables to their diet. They do not farm. A Maasai's wealth is measured by the number of cows, sheep, goats and donkeys he owns.

These handsome nomadic people live in different parts of East Africa. They are a strange mixture of two worlds, ancient and modern. Schooling is now part of a child's life. Before the practice was outlawed, the men could not marry until they had taken part in a traditional lion hunt armed only with a spear and shield. The mane was prized as a ceremonial headdress. Even today when lions see a Maasai they run for cover!

A simple but strict life is the code of the Maasai. Each day they walk many miles as their ambling herds graze alongside wild animals on the dusty terrain. They know the land well. Every rise and fall of it. Every water hole. Every marsh. Every clump of bushes and every tree. For 200 years they have lived in and around the crater. Though the name "Korongoro" is of uncertain origin, the Maasai had adopted it as part of their language before Europeans modified it to Ngorongoro.

our safari continues

As we drive along the rim, a surprise is in store at every turn. Giraffes gallop by as if floating on air. These walking "watch towers" stand about 16 feet tall, a good size for reaching their favorite food, the leaves of the acacia tree. But giraffes have to spread their slender forelegs and *stretch* down to drink.

The Maasai say that they can always tell a dominant giraffe bull: He holds his head higher than any other giraffe. And when he passes by, they say, the bulls of lower rank turn their heads as a sign of respect.

Soon your eye will become used to moving objects that suddenly turn into animals! A handsome buck impala might spring out of a bush and bound across our path in great 20-foot jumps. He may be followed by his harem, the many wives he has won in a fight with another buck, or—he may be following them.

The clowns are the warthogs, trotting like little soldiers in single file, their tails straight as rifles. Their eyes are on the top of their heads, making it convenient to watch for an enemy in all directions. When grazing they kneel on their forelegs and rip the ground for roots with their long upper tusks. Sometimes you will see a warthog with upper tusks that are a foot long.

Although animals are used to cars, it is unwise to drive too close to them. Our guide takes us to a place on the rim where we can safely get out of the car to look into the wide, open bowl. The view is breathtaking. From a height of 2,000 feet the whole floor opens up. Thick herds of game look like teeming ants roaming the plains.

driving down the crater wall

These steep forested slopes that drop down 2,000 feet make driving as exciting as a roller-coaster ride. The road is rough, narrow and dangerous. Because of sharp turns, there are times when it looks as if you are going to take a nose dive and somersault in the air! Changes in atmosphere inside the rim can create different mirages. These tricks played on the eye are distorted images of far objects. Blue and white clouds that hang low look like fluffy cotton candy drifting from the sky. Pools of water on the floor, which appear from nowhere, seem to be sheets of floating ice.

From the high slopes thousands of pink flamingoes appear in a long shimmering line, wading along the banks of Lake Makat. This white soda lake is full of alkaline water. It contains no life except tiny organisms, such as diatoms on which the flamingoes feed. When flamingoes take off for a brief flight, they explode into soft pink clouds and disappear. Soon you hear them rattle, gabble and honk on their return.

entering the crater floor

It is like a dream from a vanished world to see such a grand parade of wildlife. Herds of animals spread out over the lush crater floor. Some of the animals are frolicking, their hoofs trample the ground. Others are grazing or dozing. Their ears and tails seem to twitch constantly. Beneath the blue dome of the sky, the air is filled with a symphony of sounds: grunting and snorting, bleating and honking, whinnying and braying.

We think of animals as being "free," but from the moment of birth they live in a constant state of alertness. Each animal is deeply involved in its own daily struggle for survival. Grazing animals can tell when a predator is not actually hunting, and during these times they will feed peacefully.

Zebras fare well on the coarse tall grasses. Their grazing and trampling make the pasture ready for the wildebeests, who crop it even shorter. The gazelles follow and feed on the protein-rich young shoots. In this way animals with different needs are able to live in harmony on the same pastureland.

BURCHELL'S ZEBRAS The air vibrates with the shrill, braying barks of these wild horses of Africa as they quench their thirst at the waterhole. The young stay close to their mothers.

Zebras have close-knit families. Mares and their foals are controlled by a stallion. He will fight fiercely to protect them. Often the families join together in bands, sometimes numbering in the hundreds. As the zebras graze, at least one male stands guard to warn the others of danger.

At dawn or dusk, and in the moonlight, zebras are almost invisible. But in clear sunlight their sleek black-and-white coats stand out like stark modern paintings, each of a different design.

/ 20   WILDEBEESTS   Gray masses of wildebeest herds cluster around the waterhole. Their mellow honks can drown out all other voices. These strange looking animals, found only in Africa, have buffalo-shaped horns, long white beards and floppy manes.

Particularly during the mating season, wildebeest bulls struggle to become masters of territories so that they can attract females. A bull who already has about ten cows marks his territory with scent secreted from special glands near his eyes. Then he runs around in circles keeping his herd together and watching for intruders. He must be prepared for a ritual challenge from another bull. When faced with the challenger, he sometimes prances about kicking his hind legs, pawing the ground, snorting and grunting, and probing his horns into the earth. He whirls, lashing his tail, then drops on his knees to the ground. The bulls clash horns and spar. If the challenger fails to win the territory, he will probably end his days in an all-male herd with no chance to breed. This is nature's way of making sure that only the strongest wildebeest bulls become fathers.

During the peak calving season, which usually falls in February, all the cows give birth within the short period of a few weeks. Many young calves fall prey to such predators as lions and spotted hyenas. Yet wildebeests outnumber all other animals in the crater, by more than two to one.

THE GAZELLES You cannot miss the pale, sandy-colored ballerinas that throng the crater as far as the eye can see. They leap and frolic in graceful bounds, chasing each other throughout the day.

They look alike at a distance, but small differences tell them apart. The Thompson or Tommy has well-marked side stripes, and is smaller. It wags its tiny tail constantly, reminding you of a mechanical windshield wiper! It feeds on grasses.

The Grant's gazelle has a lighter coat and magnificent horns curving upward. It browses on small bushes and herbs.

Sometimes you will see bands of bucks roaming in bachelor herds. The does remain together. A Tommy doe will leave her herd to give birth in privacy. When the fawn is born, she licks it all over with care.

ELANDS   During the day herds of this majestic animal wander along the crater floor, grazing. They frequent the small Laiyanai forest.

A bull looks up, his eyes alert. Suddenly he leaps over the plains. Although he weighs close to 2,000 pounds, he has a dancer's grace. There is no danger—the huge eland seems to be leaping in play.

Elands are easily recognized by their heavy twisted horns and by the large dewlap hanging down from the lower neck. They have a coat of grayish-brown with light stripes.

CAPE BUFFALOS   A band of old bulls enjoys splashing and wallowing in a mud hole. Others graze nearby. First they chew the grass and swallow it. Then the grass is returned to the mouth from the stomach and is chewed again before it is digested. Buffalo spend hours grazing. Their only companions are little birds that eat insects off the buffalos' huge backs.

visiting the lerai forest

Lerai is a Maasai name for the yellow color of the fever trees in this patch of forest. The swamps, pools and marshes are fed by streams that flow down from the rim, cutting deep gorges in the crater wall. Many animals and birds live in this paradise.

ELEPHANTS Deep in the marshes are small herds of elephants that come out to graze on the surrounding plains. They walk about with a strange stillness as if they were wearing soft-soled shoes. But when disturbed or on the move, their shrill trumpeting sounds can make the forest shiver.

A grown elephant can eat from 300 to 500 pounds of food a day. When they dine, branches snap off like pistol shots cracking in the air. Their tusks, which are teeth that grow as long as the elephant lives, are used to rip bark off the trees. Their large ears act as a cooling system, flapping back and forth to create a breeze in the still air.

In this female-dominated society, the oldest cow is the leader. She is always alert, ready to give the word for the herd to move. Elephants communicate with each other through the rumbles of their stomachs, flapping of ears, the positions of their trunks, and trumpeting sounds. In the evenings, they trek to a water hole in single file for a cool dip and drink. When they suck up gallons of water, they create a cacophony of sounds.

Two young bull elephants are amusing themselves in a playful battle. One seems to decide he's had enough and walks off. The other wallows in a mud bath. He refreshes himself by drawing the watery mud up through his trunk and spraying himself with it. Teenage bulls will fight to remain with the herd, but they will eventually be driven out since there is no place for them in this female society.

The female elephants have a remarkably close family life. When a calf is too young to fend for itself, mother is always close by. This mother seems to be helping her calf to use its trunk. An elephant's trunk has over 40,000 muscles. It can stun a lion with a single blow—or pick up a single berry. If a poisonous snake bites the tip of an elephant's trunk, it can spell disaster. An elephant depends on its trunk for eating, drinking, communicating, caring for the young, lifting weights and supporting the sick or wounded members of the herd.

BABOONS Stop for a minute and watch a baboon troop on its morning stroll. Life bustles among these nimble noisy primates.

Several strong males dominate this society. They pace rapidly back and forth giving deep-throated grunts. When baboons are angry, their hairs bristle and they open their mouths to show razor-sharp teeth. If a fight threatens, one of the strong males may slap the ground as if to say, "Stop that nonsense this very minute."

Baboons sleep in trees at night to be safe from lions and leopards. During the day they are active on the ground foraging for fruit, grass, roots and insects, all part of their diet. Occasionally they kill small animals.

Grooming is an important part of baboon life. Baboons take turns grooming each other by the hour, combing the fur and picking off insects, which they eat. This is a way baboons have of showing that they care for each other.

/ 30   VERVET MONKEYS These bright creatures are cousins of the baboon. Babies stay this close to their mothers most of the time. They nurse whenever they are hungry. The older male is receiving the most attention here, as mother grooms him carefully. Later, mother may groom the baby. And she'll be groomed by other monkeys too, before the day has ended.

RHINOS Rhinos usually attack only when taken by surprise or wounded. But sometimes, they will charge at the slightest disturbing sound or smell, puffing like a locomotive at a thundering speed of 45 miles an hour. They can toss a man in the air or seriously damage an automobile with their long front horn. This horn isn't a true bone, it is made of closely packed fibers.

These prehistoric-looking animals, more accurately called "rhinoceros," are solitary by nature and are frequently alone or with their young. They can stand staring into space for hours.

COKE'S HARTEBEESTS Swift of foot, these animals depend on their speed and alertness to survive. They can run for a long time with a swinging gait. These antelopes are better known by their Kiswahili name, "Kongoni."

DEFASSA WATERBUCKS Look carefully in the swamps and you may see a handsome waterbuck with ringed horns enjoying a quiet morning near his harem and their young. If he sees you he will disappear in the bushes or swim to safety.

As their name suggests, waterbucks live near swamps, marshes and grass flats. The Maasai call them "the one with the white bottom."

THE DAINTY DIK DIK
sharp, whistling sound of t
it is frightened. When br
every few seconds with qu
The dik dik stands about
weighs less than 10 poun
falls prey to jackals and l

All Maasai children kn
shy dik dik. Long ago, du
a dik dik had to decide h
little water remaining. T
the lion most, said that t
should be allowed to dri
so angered the other ani
dik dik, leaping as thoug
invisible springs, disapp
forest. He has been hidi

/ 34

HIPPOS   No one really knows how the hippos came to the crater. Some experts believe they could have migrated, walking long distances in the night to avoid the hot sun which makes their skin glisten with red oily drops as if they were "sweating blood." These submarine-like creatures were named hippopotamus, or "river horse" by the Greeks, because they are born and live in the water. They swim with surprising ease. Only when it is cool, usually at night, do they leave their watery homes to feed. They eat as much as 200 pounds of grass, herbs and leaves in a single night. A grown male may weigh 8,000 pounds—but much of this bulk is pure muscle.

When hippos rise briefly for air you can catch a glimpse of their enormous heads, bulging eyes and bristly moustaches. They may open their mouths wide in what looks to us like a yawn. Then they disappear under water, gliding weightlessly in a circle of shimmering ripples.

A PRIDE OF LIONS In the ravines of shady trees a handsome lion sporting a black mane glances lazily at some of his pride. Lionesses with their young are panting in the heat. The lion is content. He bellows a deafening roar as if to say, "This is my territory!" He has marked it by spraying the bushes with his urine so nomadic lions will not enter.

A lion has left the pride with one of the lionesses for a few days. They look to us like honeymooners. After mating, they will return lean and hungry. When the lioness has her cubs, they become part of this large, growing family. Lion cubs are among the most helpless of animal babies. They stay with their mothers long after they are weaned.

In the midday heat there is very little activity. Invisible flames seem to scorch the grassland. The lion yawns, rolls over and stretches. He welcomes a gentle breeze that whispers through the tall swaying grasses, then ambles along to patrol his territory. The heat tires him. He drops limply to the ground near one of the lionesses and greets her by rubbing cheeks. A lioness and her young drink in a nearby stream. A mother feeds her babies. Then the adults take a snooze while the cubs play nearby.

Toward evening the lionesses become restless. It is time to hunt. They hide the cubs before leaving to search the plains. Their sharp eyes are fixed on the distance. Every muscle is stretched tightly. Their firm bodies are toned for the race. The tips of their tails twitch as they move a step at a time. Stalking carefully, they fan out on a broad "cooperative" hunt. Each one plays an important role.

Grazing animals can sense when predators are stalking. The herds look up every few seconds, eyes alert, bodies poised for flight. They know that a sudden stillness, an unfamiliar step, or even a rustle in the bush can mean sudden death if they don't react with trigger-quick speed.

Tension is mounting between the hunters and the prey. The herds move closer together, depending on each other for protection. Each time they look up, the stalking lionesses freeze, hugging the ground. The herds start to run. The lionesses watch for the weak, slow or injured to fall behind. When they see one they signal each other. In swift, effortless movements they spring out in the open. A stampede begins! Wildebeests honk loudly, bounding away.

Zebras follow with yelping barks, galloping at full speed. Close at their heels are herds of gazelle. Suddenly there is stillness in the air. The animals are quiet and begin to graze again. They know that the lionesses have captured their prey.

The hungry pride comes to the kill. At this banquet appetites are hard to satisfy. The dominant male eats first. The lionesses follow, then the cubs. The pride is not generous with their young, and if the male isn't around, he won't be summoned to the feast! After a big kill the pride spends most of the time sleeping with bellies full. Mothers become tender with their cubs and lick them lovingly all over. A big feast will keep the pride satisfied for three or four days. Then it will be time for another hunt.

The seemingly harsh law of nature that requires wild animals to prey on one another helps keep a balance among different species and serves to weed out the dull and the weak, leaving only the alert and the strong. If all animals were allowed to live, the forest and grasslands would suffer from overpopulation, which in turn would mean starvation for great numbers.

THE SCAVENGERS During the feast there is a noisy flapping from a long-legged, black-winged Marabou stork. He claps the two halves of his bill, pacing back and forth as he awaits his share of the kill. Vultures try to chase the stork away with the big capes of their wings spread out in angry threats.

A spotted hyena, recognizable by his scruffy spotted coat and rounded ears, and a jackal, with pointed ears and snout, wait nearby. There will not be enough food for all the hungry scavengers.

The jackal makes a daring move and snatches a piece of the kill from the lion. He makes off with the prize with an excited cry of "eee-eee-eeouww!" The lion lets out an angry roar and drags the rest of the kill to a safer place.

After the pride has moved away, the onlookers begin to eat the remains of the feast. The hyena has powerful jaws; she is able to eat and gain strength from the bones of a kill. When the scavengers are finished, not a sign of the kill will be seen.

These animals keep the Crater free of litter. If it weren't for them, the decaying remains of kills would soon lead to widespread sickness and death. They are enormously important to the health and safety of every animal in Ngorongoro. To us, they are perhaps less appealing than the "noble" lion. And yet, at night, when lions hear the whoop and laugh of hyenas, they will run to the sound. It means a free meal, for hyenas are excellent hunters. And lions have scavenging habits too!

birds are everywhere

Wherever you look, birds are on the move. They glide from the sky into the tall grassland, where they disappear from sight. The swamps and lakes are filled with flocks of sacred ibis, egrets, herons, yellow bill storks, crested cranes and flamingoes. In the months of January and February European storks migrate to the crater by the thousands. It is an incredible sight. Many of these birds look like fashion models parading with an air of grace and dignity.

Crested crane

Hornbill

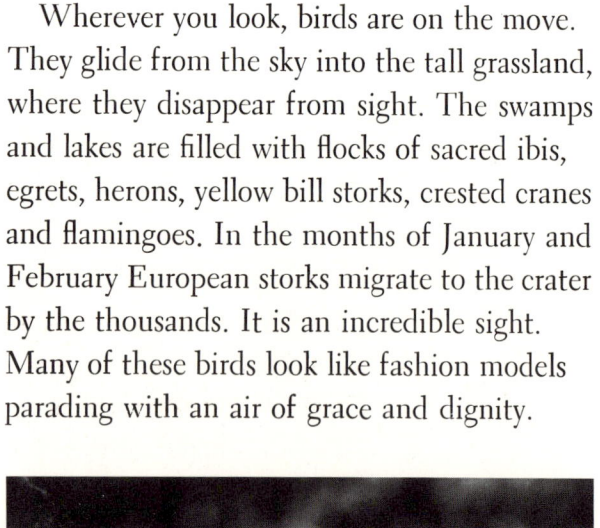

Great white heron

Yellow-billed stork

The crater also has its clumsy prancers—the seven-foot-tall ostriches. They walk about on their spindly legs and heavy bodies, swaying as they move. Though these flightless birds may wobble on their two-toed feet, they can easily outdistance an enemy, racing at a speed of about 45 miles an hour. If attacked, they use their powerful feet as a weapon.

Each bird has its own courtship dance. When a male ostrich is courting, he sings a special song. He fluffs his black feathers, tipped with white plumes on his tail, then springs off the ground in whirling movements. The gray-feathered female responds with her own dance.

Ostrich

Some birds are named simply by the way they look. The secretary bird has a crest of feathers on its head resembling ancient quill pens stuck over the ear. The hammerkop, or hammerhead, is a brown heron-like bird, whose profile is similar to the head of a hammer.

Secretary bird (above)   Hammerkop (below)

a leopard at sunset

The mood of the crater changes swiftly at sunset. Glowing orange, pink and purple colors spill across the sky and bathe the rolling grassy slopes. The crater shimmers in the dancing light. A golden-winged sunbird pours out a song in clear, lilting notes. Predators are roaming the hills and plains. Their presence is always unpredictable. A leopard, usually found on the crater rim, is hunting for his supper in the thickly wooded Lerai forest. A baboon kill would be ideal, and he is certain to find one there.

Sly and skillful in his hunting habits, he prowls silently on padded feet. He hides in a tree, sprawled out on a heavy branch covered with leaves. The sun catches his brilliant, spotted coat. His tail flicks back and forth. His cool green eyes focus on the ground where baboons are feeding and grooming. From time to time he twitches his ears to drive off buzzing insects. He is patient, waiting for the right moment.

Suddenly he springs. The baboons break out in a storm of frightened protest, but their high-pitched barking is in vain. The leopard makes a quick kill, digging in his sharp long claws and teeth. With powerful muscles he drags his prey from the ground. He anchors it high in a tree fork where hyenas and jackals cannot intrude. He dines alone.

Little is known about this solitary night creature whom the Maasai call "the prowling companion of the moon." Although leopards do not live together the male and female mate for life. Cubs remain with their mothers for from 16 to 18 months before they are able to hunt alone.

camping in the crater

Spending a night in this wild kingdom close to a host of animals is a rare and frightening experience. Darkness falls quickly and softly. The temperature drops sharply. The tent must be pitched before nightfall. Soon a bright moon creeps up in the dark sky and hangs like a lamp among sparkling stars.

Herds grazing near the tent are not disturbed by the campfire. They keep a watchful silence in the pitch black night. The breeze brings distant sounds into the open. The forest comes alive with the trumpet of elephants, the whoop and laugh of hyenas, the soft grunting of baboons, the low deep voices of hippos and the snorting of rhinos. The most fearful sound is the full-throated roar of a lion, probably a mile away but possibly right outside the tent!

A tree hyrax, no bigger than a rabbit, makes a noise that sounds like a creaking door that needs oiling. The babbling of a nightjar, a little bird with a powerful voice, wakes you up with its shrill cry: "pee-oo-wee! pee-oo-wee!" You may hear the whistling of a bat-eared fox calling out to its young. You will certainly hear the urgent cry of the hunted and the hunter; this is a time when many animals prey.

The night is long when you are not used to the strange sounds of animals. But birds soon herald the dawn with their lyrical notes. It is the hour to break camp. When the sun pours its rays over the entire crater floor, birds bask in its warm light, fluffing their wings and chirping. Grazing animals nearby look on with curious interest, honking and braying, as if to say that they have seen many tents pitched over the years under this big fig tree.

You will want to look back when driving up the slopes of the crater wall, just to take a last glimpse of this world unto itself, removed from the twentieth century. Once most of Africa was an untouched wilderness teeming with wildlife. As you leave, a Maasai in the distance calls out in that familiar greeting, "*Jambo! Jambo!*" A new day begins.

wildlife chart

Only the residents frequently seen in the Crater and on the rim are listed here. About 25,000 animals and 10,000 to 500,000 birds (depending on the number of flamingoes) live in the bowl and on the rim.

CARNIVORES *(meat eaters)*

Leopard	Black-backed jackal
Lion	Golden jackal
Bat-eared fox	Spotted hyena

HERBIVORES *(plant eaters)*

Black rhinoceros	Impala
Bushbuck	Maasai giraffe
Cape buffalo	Thompson's gazelle
Coke's hartebeest	Tree hyrax
Dik dik	Warthog
Eland	Waterbuck
Elephant	Wildebeest
Grant's gazelle	Burchell's zebra
Hippopotamus	

OMNIVORES *(meat and plant eaters)*

Baboon (primarily herbivorous)   Vervet monkey (primarily herbivorous)

BIRDS

Bustard	Flamingo	Hornbill	Plover
Crane	Guinea fowl	Ibis	Secretary bird
Eagle	Hammerkop	Nightjar	Stork
Egret	Heron	Ostrich	Vulture

index

Baboon, 29, 45
Birds, 42–43, 46

Cape Buffalo, 24
Coke's Hartebeest, 31
Crane, 42

Defassa Waterbuck, 31
Dik Dik, 32

Egret, 42
Eland, 23
Elephant, 25–28

Flamingo, 17, 42

Gazelle, 18, 22
Giraffe, 12

Hammerkop, 43
Heron, 42
Hippopotamus, 34–35
Hyena, 21, 40, 41

Ibis, 42
Impala, 13

Jackal, 40

Kiswahili, 9, 31
Kongoni, 31

Laiyanai Forest, 3, 23
Leopard, 44–45
Lerai Forest, 3, 25
Lion, 21, 32, 36–39, 40, 41
Lion hunt, 11

Maasai, 8–11, 25, 31, 32, 45, 46
Makat, Lake, 3, 17
Marabou Stork, 40
Monkey. *See* Vervet Monkey.
Montaine High Forest, 8

Ngorongoro Crater, 3, 16
   camping in, 46
   naming of, 11
Nightjar, 46

Ostrich, 43

Rhinoceros, 33

Secretary Bird, 43
Stork, 40, 42

Tree Hyrax, 46

Vervet Monkey, 30
Vulture, 40

Warthog, 14
Wildebeest, 18, 20–21, 38

Zebra, 18, 19, 39

AFRICA

EQUATOR

Lake Victoria

Ngorongoro Crater

TANZANIA